LOBELIA GARDENING HORTICULTURISTS GUIDE FROM CULTIVATION TILL COMMMERCIAL SUCCESS

The Comprehensive Handbook for Cultivation, Maintenance, Expert Tips for Blooming Success from Seed to Market

MANUEL SHELTON

Copyright © 2024 by Manuel Shelton

All rights reserved. No part of this publication may be reproduced, distributed, or transmitted in any form or by any means, including photocopying, recording, or other electronic or mechanical methods, without the prior written permission of the publisher, except in the case of brief quotations embodied in critical reviews and certain other noncommercial uses permitted by copyright law.

Disclaimer

The views expressed in this book are solely those of the author and do not necessarily reflect the views of any company, organization, or individual.

The author is not engaged in any endorsement deals or partnerships with any entities mentioned in this book. Any references to products, services, or organizations are for informational purposes only and do not constitute endorsement.

Readers are encouraged to conduct their own research and exercise their own judgment before making any decisions based on the information provided in this book

Contents

CHAPTER ONE 15
SETTLEMENTS FOR LOBELIA GARDENING............ 15
- Choosing An Ideal Site For Your Lobelia Garden . 15
- Getting The Soil Ready For Maximum Growth..... 17
- Selecting The Right Time To Plant 19

CHAPTER TWO 23
SEED PLANTING FOR LOBELIA 23
- An Illustrated Manual For Sowing Lobelia Seeds . 23
- Some Advice For Relocating Lobelia Seedlings.... 25
- Ensuring Depth And Proper Spacing 28

CHAPTER THREE... 31
LOBELIA UPKEEP AND CARE................................ 31
- Sufficient Watering For Optimal Lobelia Plant Health 31
- Methods And Schedules For Fertilization 33
- Deadheading And Pruning For Constant Blooms . 34

CHAPTER FOUR 37
MANAGING DISEASES AND PESTS 37
- Recognizing Typical Insect Pests And Illnesses Impacting Lobelia.. 37

Methods Of Controlling Chemicals And Organic Materials 38

Preventive Steps To Maintain The Health Of Your Plants 40

CHAPTER FIVE 43

EXPANDING LOBELIA 43

Techniques For Propagation: Cuttings, Division, And Seeds 43

Conditions And Timing For Effective Propagation 45

Solving Typical Propagation Problems 46

CHAPTER SIX 49

VISUAL DESIGNS WITH LOBELIA 49

Lobelia's Use In Landscape Architecture 49

Ideas For Stunning Displays With Companion Plants 50

Container Gardens: Including Lobelia 51

CHAPTER SEVEN 55

CUTTING AND MAINTAINING LOBELIA 55

Knowing When To Gather Lobelia Flowers And How To Do It 55

Methods For Preserving And Drying Lobelia 56

Innovative Applications For Harvested Lobelia 57

CHAPTER EIGHT 59

FAQS AND REGULAR QUESTIONS 59

Commonly Asked Questions 59

Taking Care Of Typical Problems And Troubleshooting Advice 61

Professional Guidance For Overcoming Obstacles 63

CONCERNING THIS BOOK

"Lobelia Gardening" is more than simply a book; it's a vital resource that reveals the techniques for growing and caring for one of the most captivating botanical jewels in nature, suitable for beginners as well as experienced gardeners. Readers are ushered into the fascinating world of lobelia plants in the introduction, where they will learn about both their extraordinary adaptability and natural beauty. As the book explores the many varieties of lobelia—each with its own distinct appeal and qualities—it becomes clear why growing lobelia is beneficial.

With "Getting Started with Lobelia Gardening," readers are given the tools necessary to design the ideal setting for their lobelia garden, marking the true beginning of the art of lobelia gardening.

Every stage is carefully laid out to guarantee success right from the start, from choosing the best spot to getting the soil ready for growth. The book offers a thorough guide with step-by-step directions and priceless advice for guaranteeing robust development and bright blooms, whether planting seeds or transplants.

A healthy garden requires care and upkeep, as any seasoned gardener is aware, and lobelia is no exception. The "Lobelia Care and Maintenance" section delves further into fertilization methods, pruning procedures, and watering regimens to make sure your lobelia plants stay vibrant and healthy all through the growing season. In addition, the book gives readers the skills necessary to recognize and repel common pests and illnesses, providing both chemical and organic management options and enabling them to protect their priceless plants.

"Propagating Lobelia" offers invaluable advice on propagation strategies and typical problem-solving for people who are keen to grow their lobelia garden or share its beauty with others. In the meanwhile, "Designing with Lobelia" provides imaginative ideas for combining these lovely plants into container gardens, companion planting schemes, and landscape designs.

A successful gardening trip culminates in the harvesting and preserving of lobelia flowers, and "Harvesting and Preserving Lobelia" offers priceless insights into the finest procedures for guaranteeing a plentiful harvest and extending the beauty of lobelia blossoms.

For aficionados looking to realize the full potential of these fascinating plants, "Lobelia Gardening" is the go-to reference since it provides expert advice, troubleshooting suggestions, and answers to commonly asked concerns.

Overview of Gardening using Lobelias

Lobelia plants are a global favorite among gardeners due to their delicate blossoms and vivid colors. Anyone hoping to add a touch of elegance to their landscape or garden must appreciate the beauty and adaptability of lobelia plants. These pretty flowers will liven up any outdoor area. They come in a variety of blue, purple, pink, and white hues.

Realising The Versatility And Beauty Of Lobelia Plants

As members of the Campanulaceae family, lobelia plants are indigenous to Australia, Africa, and both North and South America. They are renowned for having tiny, star-shaped flowers that bloom lustfully all summer long, producing a stunning show of color. The propensity of lobelia to cascade over edges is one of its most appealing qualities, which makes them perfect for hanging baskets, window boxes, and other containers.

Lobelia plants are not only aesthetically pleasing but also provide gardeners with a number of useful advantages. They require very little upkeep and can grow in a variety of environments, including places that are partly shaded and sunny. Because of their adaptability, they can be used for a wide range of landscaping tasks, including ground cover, filler plants, and borders and edging.

Advantages of Lobelia Planting

The potential of lobelia to draw pollinators to the garden, including bees, butterflies, and hummingbirds, is one of the main advantages of planting them. These helpful insects rely on the nectar-rich blooms as a valuable food source, which supports nearby ecosystems and fosters biodiversity. Lobelia plants are a natural and environmentally beneficial way to control pests like aphids and whiteflies, in addition to drawing pollinators.

Growing lobelia also has the advantage of being versatile in garden design. Lobelia plants work well in a wide range of landscaping designs, whether you're designing an elegant flower bed or a relaxed cottage garden. They are perfect for use as ground covers, border plants, or filler plants in mixed flower arrangements because of their small size and trailing habit.

Synopsis of Lobelia Types

There are many different types of lobelia plants, and each has special traits and needs for growth. Among the most common varieties of lobelia are:

Lobelia erinus: Often referred to as trailing lobelia, this cultivar is highly valued for both its abundance of tiny blooms and its trailing nature. It is frequently used in window boxes, hanging baskets, and other containers where it can overflow and produce a colorful cascade.

Also referred to as the cardinal flower, Lobelia cardinalis is highly valued for its towering spikes of vivid red blossoms. It is a great option for water gardens and pond margins because it is native to North America and grows well in damp, marshy environments.

Lobelia siphilitica: Known for its tall spikes of sky-blue flowers, this type is highly valued. It is also referred to as great blue lobelia. It is a great option for shaded garden beds and borders because it is native to North America and grows well in damp, forest settings.

Lobelia laxiflora: Often referred to as Mexican lobelia, this cultivar is highly valued for its upright growth habit and vivid orange-red flowers. It is a native of Mexico and does well in sunny, well-drained soils, which makes it a great plant for sunny borders and rock gardens.

You can appreciate the beauty and adaptability of lobelia plants year-round by choosing the proper kind for your yard and creating the ideal growing environment. Gardeners can find something to love about lobelia gardening, whether it's the beauty of these beautiful flowers, attracting pollinators, or adding color to your landscape.

CHAPTER ONE

SETTLEMENTS FOR LOBELIA GARDENING

Choosing An Ideal Site For Your Lobelia Garden

Selecting the ideal location for your lobelia garden establishes the conditions for robust development and vivid blossoms.

Look for places in your garden that receive four to six hours of sunlight per day as lobelias-like settings with partial shade to full sun. Steer clear of extremely shaded areas as these can result in lanky growth and fewer blossoms.

Lobelia likes its soil to be healthy, well-draining, and pH neutral to slightly acidic. Make sure the place has adequate drainage because Lobelia dislikes to sit in soggy soil. Take this into consideration while selecting a location.

To enhance the texture and drainage of your hard or compacted soil, mix it with organic matter like compost or peat moss.

Additionally, consider the surrounding surroundings. Since Lobelia is susceptible to intense heat, pick a spot that gets some afternoon sun protection if you live in an area with scorching summers. If you live in a colder region, on the other hand, pick a location with lots of sunlight to promote strong development and blossoming.

Finally, when choosing the place, take aesthetics and accessibility into account. Plant your lobelia in a place where you can easily appreciate its beauty, such as in a mixed container garden, along a pathway, or in a hanging basket.

Consider how its color, height, and texture will work well with the other plants in your garden as well.

Getting The Soil Ready For Maximum Growth

A vital first step in making sure your lobelia garden takes off is preparing the soil. Clear the planting area of any rocks, weeds, or rubbish first. By removing these impediments from the soil, you can avoid competition for nutrients and facilitate the lobelia's establishment.

Next, use a garden fork or tiller to loosen the soil until it is about 6 to 8 inches deep. Cultivation is the loosening process that promotes healthy development by improving soil aeration and facilitating root penetration. Take care not to overwork the soil, as this might damage the beneficial microbes and soil structure.

It's time to amend the soil with organic matter to improve soil texture and add nutrients after it has been loosened up.

Add a good helping of compost, well-rotted manure, or peat moss to the soil and thoroughly mix to make sure it is evenly distributed. The addition of organic matter improves soil drainage, fertility, and moisture retention, which results in the ideal conditions for lobelia growth.

Consider doing a soil test to determine the pH and nutrient content of the amended soil. Lobelia likes soil that is between 5.5 and 7.0 pH, which is slightly acidic as opposed to neutral.

If required, use amendments to raise or lower pH, such as elemental sulfur to raise pH, in accordance with the suggestions derived from the findings of your soil test.

Lastly, use a rake to level the soil top to prepare the planting bed for your lobelia seeds or transplants. A thorough soil preparation process lays the groundwork for robust development and a profusion of blooms throughout the growing season.

Selecting The Right Time To Plant

When planting lobelia, timing is essential to a successful establishment and healthy growth. In most areas, lobelia is grown as an annual, yet in warmer temperatures, certain species may act more like perennials. The ideal planting time will be determined by you after you have a better understanding of the local climate and the requirements unique to the lobelia variety you are cultivating.

It is generally advised to grow lobelia in temperate climates with distinct seasons when the risk of frost has gone in the spring. This guarantees that the young plants won't be harmed by the cold, enabling them to take root and grow quickly. On the other hand, lobelia can be planted in the autumn for winter and spring blooms if it's being grown as a winter annual in a mild area.

Make sure the soil temperature is warm enough for lobelia development before planting. The ideal range of soil temperatures for lobelia to germinate and establish roots is between 55°F and 65°F (13°C and 18°C). Measure the temperature at a depth of 2 to 4 inches with a soil thermometer to see if the conditions are right for planting.

If you are beginning lobelia from seed, follow the suggested spacing and depth listed on the seed packet and plant the seeds straight into the prepared soil. An alternative is to start seeds indoors eight to ten weeks prior to the final anticipated date of frost, then move the plants outside after they have a few genuine leaves.

Those who like their gardens transplanted, buy healthy lobelia seedlings from a garden centre or nursery. Plant the seedlings at the same depth as they were growing in their containers, being careful not to damage their fragile roots throughout the transplanting process. After planting, give the transplants plenty of water to help them acclimatize to their new surroundings.

You can give your lobelia the best chance of thriving and embellishing your garden with its lovely blossoms by selecting the right planting time and using the right planting practices.

CHAPTER TWO

SEED PLANTING FOR LOBELIA

An Illustrated Manual For Sowing Lobelia Seeds

The procedure of planting lobelia seeds is easy to follow and only requires a few basic steps. Here is a starting point guide to assist you:

First things first, get your lobelia seeds planted in a well-draining, healthy soil mixture. In the area, you want to plant, loosen the soil and pull weeds and rubbish.

The ideal site for your lobelia plants is one that receives enough sunlight throughout the day. Lobelia plants prefer partial shade to full sun. Make sure there is some shade from the scorching afternoon sun, particularly in warmer climates.

Plant the seeds: Distribute the lobelia seeds uniformly across the surface of the prepared soil.

Given their small size, lobelia seeds should not be buried too deeply. To make sure there is adequate seed-to-soil contact, just gently press them into the ground with your fingertips.

Water sparingly: After the seeds are sown, lightly sprinkle the area with a hose or use a watering can. Steer clear of soaking the soil, as this may push the seeds around or let them decay.

Sustain moisture: Throughout the germination stage, keep the soil continuously damp but not soggy. Based on the local weather, you might need to water lightly every day or every other day.

If necessary, thin seedlings: To ensure appropriate spacing, you may need to thin out the seedlings as they emerge and form their first true leaves.

To provide sufficient airflow and growth, give each plant a gap of around 6 to 8 inches.

Fertilise sparingly: While intensive feeding is not necessary for Lobelia plants, you can promote healthy growth and bloom by using a balanced liquid fertilizer once every two to three weeks during the growing season.

Savor your blooms: Depending on the variety, your lobelia plants should start to bloom in a few weeks to months with the right care and attention. Enjoy the lovely arrangement of delicate flowers in your yard while you kick back and unwind!

Some Advice For Relocating Lobelia Seedlings

To guarantee the survival and continuous growth of lobelia seedlings, transplanting them is a delicate procedure that needs to be handled carefully. The following advice will help you effectively transplant your lobelia seedlings:

Select the ideal moment to transplant your lobelia seedlings into their permanent growing location: Do not do so until your seedlings have produced at least two pairs of genuine leaves. This typically occurs four to six weeks following germination.

The new planting location should be prepared by loosening the soil and adding organic materials, such as compost or well-rotted manure, to promote drainage and fertility before transplanting.

Water the seedlings: To help break up the soil around their roots and lessen transplant shock, give the seedlings plenty of water a few hours before transplanting.

Handle carefully: To prevent breaking the fragile stems, use care while taking the seedlings out of their trays or containers by holding them by the leaves or root ball.

Create planting holes: Create holes at the newly designated planting location that are marginally bigger than the seedlings' root balls.

The recommended spacing for the lobelia variety you are growing should be followed when spacing the holes.

Planting at the proper depth requires that each seedling be planted at the same depth as it was in its prior container. To guarantee strong touch and stability, carefully press the earth around the roots.

Thoroughly water: After transplanting, give the seedlings enough water to let the soil surrounding their roots settle and to provide them the moisture they require to become established in their new home.

Continue to water the transplanted seedlings on a regular basis, making sure the soil is always damp but not soggy until they establish themselves and begin to grow quickly.

Ensuring Depth And Proper Spacing

When planting lobelia seeds or transplanting seedlings, proper spacing and depth are crucial considerations to make sure the plant grows and develops to its full potential. This is how to make sure you do it correctly:

Planting lobelia seeds or relocating seedlings requires careful spacing to ensure sufficient ventilation, solar radiation, and space for development.

As with other lobelia varieties, pay attention to the spacing guidelines recommended for your particular type. These should be between 6 and 12 inches apart.

Depth: To encourage good root development and avoid problems like damping off, it's crucial to plant lobelia at the proper depth whether planting seeds or transplanting seedlings.

Don't bury seeds too deeply; instead, gently press them into the soil's surface. Plant seedlings at the same depth that they were growing at when you transplant them, making sure the dirt covers the entire root ball.

When planting lobelia, it's important to consider both spacing and depth to give your plants the best chance of thriving during the growth season.

CHAPTER THREE

LOBELIA UPKEEP AND CARE

With its beautiful, trailing blossoms, lobelia adds a charming touch to any container, hanging basket, or garden. Adequate care and maintenance are crucial for maintaining the health and well-being of your lobelia plants. This thorough guide will help you make sure your lobelia plants thrive the entire growing season.

Sufficient Watering For Optimal Lobelia Plant Health

Lobelia plants require regular irrigation to stay healthy and vibrant, but finding the ideal balance is essential. Although constantly moist soil is preferred by lobelia, it's crucial to avoid overwatering as this can cause root rot and other problems. When watering your lobelia plants, try to water them deeply and let the soil dry a little in between.

You might need to water more regularly in the sweltering summer months when lobelia is in full bloom in order to keep the soil from drying up entirely. But exercise caution when watering too much, as wet soil might encourage the growth of fungi. Use a soaker hose or drip watering system to provide water directly to the root zone without over-wetting the leaves in order to maintain ideal moisture levels.

Make sure the drainage in pots or hanging baskets is adequate before watering lobelia plants to avoid soggy soil.

This can be accomplished by utilizing pots with drainage holes and filling the bottom of the container with gravel or perlite before planting.

Furthermore, periodically check the soil's moisture content by sticking your finger a few inches below the surface. If it feels dry, it's time to water.

Methods And Schedules For Fertilization

Supporting the growth and flowering of lobelia plants requires fertilization. The vital nutrients required for robust foliage and an abundance of flowers can be found in a well-balanced fertilizer. To create a solid base for your lobelia plants, start by adding a slow-release, balanced fertilizer to the soil when you plant.

To promote continued blooming throughout the growth season, add a water-soluble fertilizer every two to three weeks. Select a fertilizer that contains more phosphorus to encourage the growth of flowers. As an alternative, to provide the best possible nourishment, use a liquid fertilizer designed especially for flowering plants.

Pay close attention to the manufacturer's instructions when applying fertilizer to prevent overfertilization, which can cause nutrient imbalances and harm to the plants. After fertilizing, make sure to fully water the soil to help disperse the nutrients and avoid fertilizer burn.

For lobelia plants, organic alternatives like fish emulsion or compost tea can be helpful in addition to commercial fertilizers. Over time, the health of the soil is improved by these natural fertilizers, which release nutrients gradually. For long-term growth and to improve the growing environment, add organic matter to the soil once a year.

Deadheading And Pruning For Constant Blooms

Deadheading and pruning are crucial upkeep activities that assist lobelia plants remain compact and bushy-looking while extending their flowering season.

By removing wasted flowers, or deadheading, you can stop the plant from focusing its energy on producing seeds and promote the growth of new blooms.

Check your lobelia plants frequently for faded or wilted flowers, and pinch them off with your fingers or, for bigger clusters, use sterile pruning shears. Throughout the growing season, deadheading should be done to maintain the plant's appearance and encourage ongoing flowering.

Pruning on sometimes can help revitalize lobelia plants and encourage new growth in addition to deadheading. Pruning lanky or straggling stems will promote branching and result in a more compact, fuller habit.

To increase air circulation and lower the chance of fungal infections, concentrate on eliminating any harmed, diseased, or densely packed growth.

To promote new growth in the desired direction, make clean cuts directly above a set of healthy leaves or a lateral bud when pruning lobelia plants.

Removing more than one-third of a plant's total growth at once or cutting it into woody stems should be avoided since these actions might cause stress to the plant and impede its ability to heal.

You may enjoy a lush, colorful display of flowers throughout the growth season by implementing these pruning and deadheading procedures into your lobelia care regimen.

Maintaining your garden regularly not only makes it more aesthetically pleasing but also helps your lobelia plants stay healthy and vibrant for a longer period of time.

CHAPTER FOUR

MANAGING DISEASES AND PESTS

Recognizing Typical Insect Pests And Illnesses Impacting Lobelia

With its soft blossoms and rich foliage, lobelia is susceptible to a number of pests and illnesses. It's essential to recognize these problems early on if you want to keep your plants looking beautiful and healthy. Aphids are among the most frequent pests to afflict lobelia.

These microscopic insects cause the plant's leaves to droop and become yellow by sucking in its sap. Additionally, you could observe a sticky substance on the leaves called honeydew.

The whitefly, which frequently gathers on the undersides of leaves, is another prevalent pest. If left untreated, they can swiftly infest your lobelia and resemble small white moths.

Apart from pests, lobelia may also be vulnerable to several kinds of illnesses. Powdery mildew is one of the most common, appearing on the leaves and stems as a white, powdery substance.

If left untreated, this fungal illness can spread swiftly and prefers humid environments. Root rot is another problem to be aware of, which is brought about either by overwatering or poorly draining soil.

Yellowing leaves slowed development, and an unpleasant smell coming from the roots are some of the symptoms.

Methods Of Controlling Chemicals And Organic Materials

You have a few choices at your disposal for managing diseases and pests on your lobelia. Many gardeners choose to use organic methods since they are better for the environment and helpful insects. Try misting your plants with a neem oil and water mixture to get rid of aphids and whiteflies. The way neem oil works is by smothering the bugs and changing the way they feed. Additionally, you can introduce beneficial insects that eat pests in gardens like ladybirds and lacewings, which feed on aphids.

Chemical management measures can be necessary if the infestation is severe or persistent. Aphids, whiteflies, and other soft-bodied insects can be effectively controlled with horticultural oils and insecticidal soaps. To prevent leaf damage, carefully follow the instructions and steer clear of spraying on hot, sunny days. Fungicides containing sulfur or copper can help limit the generation of spores in fungal infections such as powdery mildew. Once more, to guarantee safe and efficient application, it is imperative to read and abide by the label instructions.

Preventive Steps To Maintain The Health Of Your Plants

Although managing pests and illnesses might be a burden, there are actions you can do to stop them from happening in the first place. Creating ideal growing conditions for your lobelia is one of the best strategies to maintain its health.

This entails giving them enough sunlight and ventilation as well as putting them in soil that drains effectively. Keep your plants from becoming overcrowded since this might foster an environment that is conducive to the growth of pests and illnesses.

For early identification and management, it's also critical to regularly check your plants for warning indications of problems.

In order to stop the illness from spreading, remove any afflicted leaves or stems right away. Furthermore, removing possible hiding places for pests and diseases can be achieved by maintaining proper garden hygiene by clearing away trash and fallen leaves.

To naturally repel pests, think about using companion planting strategies. Aphids and other common garden pests are known to be repelled by plants like marigolds, basil, and petunias.

Lastly, keep in mind that overfertilizing your lobelia may increase their susceptibility to illness. Choose a balanced fertilizer instead, and make sure to apply it as directed by the manufacturer.

You can maintain the health and growth of your lobelia plants for many seasons to come by practicing proactive and watchful gardening.

CHAPTER FIVE

EXPANDING LOBELIA

Techniques For Propagation: Cuttings, Division, And Seeds

Lobelia is a popular choice for enhancing the beauty of gardens and containers because of its delicate blossoms and brilliant colors. There are three ways to propagate lobelia: division, cuttings, or seeds. Each has benefits and drawbacks of its own.

Lobelia can be multiplied easily and successfully by division, especially if the plants are established and have gotten too crowded. To split lobelia, carefully remove the plant from its clumps and divide it into smaller portions, making sure that the roots remain linked to each division. Plant the divisions again in soil that drains well, and moisten them consistently until they take root in their new position.

Another dependable way of propagation is taking cuttings from lobelia; this is very useful for making sure that the traits of the parent plant are retained in the new plants. Cut a few healthy stems from the parent plant slightly below a leaf node. To promote the growth of roots, remove any lower leaves that reveal the nodes. Then, dip the cut end into the rooting hormone. After planting the cuttings in a potting mix that is damp but drains well, place them in a warm, bright area until they begin to root.

One economical method that makes it possible to generate a lot of plants very rapidly is to grow lobelia from seeds. First, plant the seeds in sterile seed-starting mix-filled trays or pots. Gently press the seeds into the soil, then sprinkle a thin coating of vermiculite or perlite over them.

To promote germination, keep the soil consistently moist and at a warm temperature of about 70–75°F (21–24°C). The seedlings can be moved into individual pots or straight into the garden once they have produced multiple sets of genuine leaves.

Conditions And Timing For Effective Propagation

Depending on the approach employed, there are differences in the conditions and timing for lobelia propagation.

Early spring or early autumn, when the weather is cool and the plants are actively growing, are the ideal times to divide. Lobelia plants can be cut down at any point during the growing season, but the best times to take cuttings are in the spring or early summer when the growth is at its fastest. You can sow seeds directly in the garden once the risk of frost has passed, or you can sow them indoors 8–10 weeks before your area's last frost date.

Whatever the method of propagation, lobelia needs specific conditions in order to develop well. Because lobelia needs a slightly acidic pH between 5.5 and 6.5, provides well-draining soil rich in organic matter. As soon as newly propagated lobelia begin to take root, water them frequently to keep the soil evenly moist but not soggy, and shield them from direct sunshine.

Solving Typical Propagation Problems

Lobelia is not the only plant for which propagation might be difficult at times. Root rot, damping-off disease, and poor germination are common problems with lobelia propagation. Scarify the seeds (gently rub between two sheets of sandpaper) or soak them in warm water for the entire night before sowing to increase germination rates.

Use sterile potting mix and pots, and refrain from overwatering recently propagated plants to prevent damping-off and root rot. Plants should be spaced adequately to allow for adequate air circulation; overcrowding can encourage fungal diseases.

Consider using a heating mat to provide bottom heat to promote root growth if your cuttings or divisions are struggling to take root or exhibit other signs of stress.

Until newly propagated plants become established, keep the humidity levels high around them by covering them with a plastic dome or putting them in a humidity tray.

You may effectively propagate lobelia and enjoy its lovely blossoms in your garden or containers with the right care and attention.

CHAPTER SIX
VISUAL DESIGNS WITH LOBELIA

Lobelia's Use In Landscape Architecture

A multipurpose and vibrant plant, lobelia may be used in a variety of ways to accentuate the beauty of your landscape design. Planting lobelia as edging plants or along borders and paths is a common technique used in landscape design. Low-growing cultivators, like Lobelia erinus, provide a carpet of vivid flowers that give garden beds and borders a splash of color.

Lobelia can also be used in landscape design by cascading down retaining walls or planting in rock gardens. The trailing types, like Lobelia pendula, soften the borders of hardscapes and give rocky regions more texture with their cascading impact. Lobelia plants in these areas serve to break up the monotony of hard surfaces and add visual interest.

Additionally, lobelia can be used as a filler plant to give your outdoor living spaces a pop of color and texture in mixed containers or hanging baskets. Choose trailing or compact lobelia cultivars for container gardening that will blend in with other plants and create a well-balanced design. For a colorful and striking arrangement, try combining lobelia with complementing flowers and foliage plants.

Ideas For Stunning Displays With Companion Plants

A gardening practice known as companion planting combines various plant types to foster mutual benefit and a peaceful environment. There are a few choices to think about when it comes to lobelia companion planting for a gorgeous show in your yard.

The trailing annual bacopa (Sutera cordata), which has little white or blue flowers, is a well-liked companion plant for lobelia. When planted in containers or hanging baskets, lobelia and bacopa produce a striking contrast of hues and textures, with the lobelia's large blooms standing out against the delicate bacopa petals.

Alyssum (Lobularia maritima), a low-growing annual with clusters of small, fragrant blooms, is another plant that goes well with lobelia.

Combining lobelia and alyssum in garden beds or borders results in a colorful carpet that draws pollinators and gives your landscape a pleasant scent.

Container Gardens: Including Lobelia

Lobelia's compact stature, trailing habit, and bright blossoms make it a great plant for container gardens. Selecting the appropriate lobelia variety and container size is crucial for creating a vibrant and healthy show when adding lobelia to container gardening.

For container gardens, choose compact or trailing lobelia cultivars such as Lobelia erinus or Lobelia pendula. These types grow nicely in hanging baskets and pots, and they create a colorful waterfall that overflows the sides of the container.

Use premium, well-draining, nutrient-rich potting mix when growing lobelia in pots. Depending on the type of lobelia you are growing, use containers with drainage holes to avoid waterlogging and root rot. Then, position the pots in a spot that receives partial to full sun.

To maintain healthy growth and continuous flowering, fertilize with a balanced liquid fertilizer every few weeks and water lobelia often to keep the soil equally moist but not soggy.

Deadhead spent flowers often to encourage reblooms and keep the arrangement neat.

Lobelia's vivid hues and graceful trailing habit lend beauty and grace to container gardens, which is why they are a favorite among gardeners who want to create eye-catching arrangements in constrained areas.

Try experimenting with different types and companion plants to make striking arrangements that will enliven your outdoor spaces.

CHAPTER SEVEN

CUTTING AND MAINTAINING LOBELIA

Knowing When To Gather Lobelia Flowers And How To Do It

A lovely accent to any garden is lobelia, with its exquisite blue, purple, or white blossoms. Timely flower harvesting is essential if you want to appreciate its beauty for an extended period of time. Harvesting lobelia blooms is usually best done in the morning after the dew has dissipated but before the sun reaches its zenith. By doing this, the flowers are guaranteed to be at their most brilliant and fresh.

Look for fully open, non-wilted blossoms while picking lobelia flowers. Using clean, sharp garden scissors, gently squeeze or cut the stem immediately below the flower head. When harvesting, take care not to injure the plant or neighboring buds. To ensure that the plant blooms continuously all season long, try to leave some blossoms on it.

Wait until the little seed pods have grown and the flowers have faded before collecting lobelia for its seeds. When these pods reach maturity, they will turn brown, signifying that the seeds are ready to be harvested. Just remove the pods from the plant and let them fully dry before storing.

Methods For Preserving And Drying Lobelia

Lophia flowers are easy to dry, so you can continue to appreciate their beauty long after the growing season is over. First, trim any extra leaves from the stems of the lobelia so that just the blooms remain before drying them. Next, tie multiple stems in a bundle and fasten it with a string or rubber band.

Place the bundles upside down in a warm, dry place with plenty of ventilation, like a covered porch or well-ventilated room. Hanging them in the middle of the day will prevent the blooms from fading. For many weeks, let the lobelia air dry; at that time, make sure it's not too dry.

Carefully remove the blossoms from the stems once they are absolutely dry and brittle to the touch. The dried flowers can then be kept out of direct sunshine and moisture by placing them in airtight containers. When properly dried, lobelia holds its color and shape for months, making it ideal for a variety of crafts and arrangements.

Innovative Applications For Harvested Lobelia

Beyond conventional floral arrangements, harvested lobelia flowers can be creatively used for many different purposes. To give your home a gentle floral aroma and a splash of color, try adding dried lobelia to potpourri blends. Dried lobelia is also a natural substitute for artificial fragrances in homemade sachets or herbal pillows.

To maintain the delicate beauty of lobelia blooms, consider pushing them between the pages of a heavy book for a more creative approach. The blooms can be used to decorate framed botanical artworks, bookmarks, and greeting cards after they have been pressed.

Making handmade bath and body products using collected lobelia is another inventive use for them. To add some color and a hint of flowery aroma to body scrubs, bath salts, or herbal teas, add dried lobelia flowers. These products will not only have a stunning appearance but also offer a comforting and fragrant experience.

You can take advantage of lobelia flowers' beauty and adaptability all year long by carefully collecting and storing them. Lobelia lends a sense of organic elegance to any project, whether it is dried for decoration or used to make handmade goods.

CHAPTER EIGHT

FAQS AND REGULAR QUESTIONS

Commonly Asked Questions

Q: How often should my lobelia plants be watered? A: Lobelia plants dislike being soggy soil; instead, they need their soil to be continually moist.

When the top inch of soil feels dry to the touch, that's the ideal time to water them. This may require watering every two to three days in hot, dry weather, or less frequently in cooler months, depending on your environment and soil conditions. To prevent wetting the leaves, which might cause illness, make sure to water the plants at their bases.

Do lobelias require direct sunlight or shade? A: Lobelias grow well in both full and partial sun. They can withstand little shade, especially in warmer locations, but they like at least six hours of sunlight per day. If you put them where the afternoon light is strong, you may assist keep them from wilting in the hot heat by giving them some afternoon shade.

How should lobelia blooms be deadheaded? A: Lobelia plants benefit from deadheading, or removing spent flowers since it prolongs their blooming season and keeps the plants from prematurely going to seed. When faded flowers begin to wilt, just pinch them off at the root. This encourages more abundant blossoming all season long in addition to keeping your plants looking neat.

Can I cultivate lobelias in a container? A: Definitely! Because lobelias grow well in containers, they are ideal for bringing color to tiny areas like patios and balconies. Select a container with well-drained holes and add premium potting mix to it. Remember that because pots dry out faster than soil, plants grown in containers could require more regular watering than those planted in the ground.

Are lobelias resistant to deer? A: Lobelias are often not at the top of deer's chosen menu, but no plant is 100% deer-proof. Compared to certain other garden plants, they are less attractive to deer because of their bitter taste and texture. Notwithstanding, deer may still nibble on lobelias, particularly new or sensitive growth, in places with high deer populations or during periods of extreme hunger. To help safeguard your lobelias, think about planting companion plants that are resistant to deer or applying deer repellents.

Taking Care Of Typical Problems And Troubleshooting Advice

Problem: Despite receiving regular watering, my lobelia plants are wilting. Solution: A number of things, such as root rot, overwatering, and underwatering, can lead to wilting. After determining the soil's moisture content, modify your watering schedule. Make sure the soil is not too wet for the plants to sit in since this could suffocate the roots. Remove any afflicted plants with care, cut off any rotted roots, and replant in new soil if root rot is detected.

Issue: My lobelia plants' foliage is becoming yellow. Solution: Symptoms of yellowing foliage include bugs, overwatering, and nutrient deficits. Review your watering techniques and make any necessary adjustments. To give your lobelias the vital nutrients they need, think about fertilizing them using a balanced, water-soluble fertilizer. Check the plants for evidence of pests that can harm the leaves, such as spider mites or aphids. Use neem oil or insecticidal soap to treat infestations as soon as possible.

The issue is that my lobelia plants aren't blooming as much as I would have liked. Solution: Inadequate sunshine, excessive crowding, or a deficiency of nutrients can all contribute to poor blooming. Make sure your lobelias get enough sunlight each day, and remove any congested plants to encourage better airflow. To promote additional blooms, think about using a fertilizer designed specifically for annual flowers in your plants. Regularly deadhead wilted flowers to encourage ongoing blooming all season long.

Professional Guidance For Overcoming Obstacles

Tip: Provide Adequate Drainage The health of lobelia plants depends on adequate drainage. If left in soggy conditions, they quickly succumb to root rot and require moist but well-drained soil. Use organic matter to promote drainage in thick clay soil before planting. Use potting mix made especially for container planting; it usually has ingredients like vermiculite or perlite to help with drainage.

Advice: Keep an Eye Out for Pests and Diseases: Frequently check your lobelia plants for symptoms of pest infestations or illnesses, such as deformed growth, yellowing leaves, or odd markings or patches. Aphids, slugs, and snails are common pests that target lobelias. In certain situations, illnesses like powdery mildew or root rot can also develop. Any problems should be promptly identified and treated to stop them from getting worse and harming your plants even more.

Recommendation: Take Into Account Companion Planting Using companion planting as a method can help your lobelia garden become more vibrant and healthy. By combining lobelias with other plants that share their cultural needs, you can increase the resilience of your garden by attracting beneficial insects and keeping pests away. Lobelias pair well with marigolds, petunias, and alyssum, which offer comparable hues and textures as well as extra advantages like insect control and pollinator support.

Try out several combinations to see what functions best in your garden.

www.ingramcontent.com/pod-product-compliance
Lightning Source LLC
Chambersburg PA
CBHW050019230526
45470CB00003B/1045